J $13.95
358   Nicholas, John
Ni   Army air support

DATE DUE

OC 5'91	NOV 18 '93		
OC 31'91	JUN 7 '94		
NO 29'91	SEP 25 '95		
AP 11'92	NOV 16 '95		
JY 16'92	AP 02 02		
JY 23'92	JY 08 '04		
SE 8'92	DE 20'04		
NO 16'92	N 29		
DE 1'92			
MR 15'93			
JY 28'93			

ARMY
AIR SUPPORT

ARMY AIR SUPPORT

by John Nicholas

Rourke Enterprises, Inc.
Vero Beach, Florida 32964

CH-47D Chinook helicopters lift a big M198 155mm howitzer, its eleven-man crew, and 32 rounds of ammunition — a combined cargo of 11 tons.

Library of Congress Cataloging-in-Publication Data
Nicholas, John, 1944-
  Army air support/by John Nicholas.
  p. cm. — (The Army library)
  Includes index.
  Summary: Describes different types of air support used by the United States Army at various times in history, including aerial surveillance, cargo transport, attack, and evacuation.
  ISBN 0-86592-421-X
  1. United States. Army — Aviation — Juvenile literature. 2. Close air support — Juvenile literature. 3. Aeronautics, Military — United States — Juvenile literature.[1. United States. 2. Close air support. 3. Aeronautics, Military.] I. Title. II. Series. Nicholas, John, 1944- Army library.
UG633.N765   1989   88-13803
358.4 - dc19   CIP
                                            AC

CONTENTS

1. Early Days   6

2. The Modern Air Arm   11

3. Spies in the Sky   18

4. Cargo Flight   26

5. Attack!   30

6. Support Duties   38

7. The Future   42

    Abbreviations   46

    Glossary   47

    Index   48

EARLY DAYS

Since the early days of aviation, the United States Army has been interested in using aircraft to observe and disarm the enemy. In the Civil War (1861-65), balloons were used to spy on enemy positions. In December 1907, just four years after the Wright Brothers first flew a powered aircraft, Brigadier-General James Allen, Chief Signal Officer of the U.S. army, offered industry the specification for a military aircraft.

On February 10, 1908, the army signed a contract for a Wright Model A biplane. For the next ten years, building an army aviation unit progressed slowly. Aviation had been controlled by the Aeronautical Division of the Army Signal Corps since August 1907, but in July 1914 control was transferred to the Aviation Section of the Army Signal Corps. By this time the army had purchased some aircraft, but by 1917 it still had less than 250 planes.

The army began its association with aircraft in 1908 when it signed a contract for a Wright Model A biplane.

It may seem surprising today to recall that the U.S. Army was the only instrument of land-based air power for the first forty years of military aviation in the United States. ▲

Lt. Thomas E Selfridge (left) and Alexander Graham Bell at
◄ *Fort Myer, Virginia, prior to a test flight with Orville Wright who, five years earlier, had made the world's first manned flight of a heavier-than-air flying machine.*

World War One, (1914-18) broke out in Europe during August 1914, and the United States joined the war in 1917. Aviation was still under the authority of the Signal Corps until May 1918, when President Woodrow Wilson established the Army Air Service. Personnel strength rose to a peak of 195,000 from just 200 three years before.

During World War One, the Army Air Service purchased several thousand planes from friendly European countries. When peace came, it had 740 aircraft in its squadrons. This was about 10 percent of the total air force fighting the Germans and the Austro-Hungarians. Nevertheless, army fighter pilots shot down more than 700 enemy planes and 73 observation balloons.

In July 1926, the Army Air Service was abolished and the Army Air Corps took its place. By this time, personnel strength had dropped to 9,000, but signs of war in Europe once more sent recruitment up. By the time that the Japanese attacked Pearl Harbor in December 1941, Army Air Service personnel totaled 152,000. During World War One almost all the aircraft flown by American units were European. World War Two (1939-1945) was to be very different.

By the end of World War Two in 1945, the United States Army had more than two million men and women in aviation. The unit's name had changed again. In June

The army was responsible for the development of giant bombers like this B-29 Superfortress used to attack Germany ▶ and Japan during World War Two.

Today's mighty air force is built on many years of development, during which the army founded the largest and strongest military air fleet in the world. ▼

1940, the Army Air Corps became the Army Air Force, a much more fitting title. Between 1941 and 1945 U.S. factories had produced nearly 300,000 planes. In that period, the Army Air Force received about 158,000 aircraft, of which around 88,000 were destroyed in the war. The navy and America's allies got the rest.

Building these planes had been a colossal effort, and American factories had installed enormous production lines that had not even reached their peak when peace came. In the last year of war alone, U.S. factories built over 110,000 planes. At that time, the U.S. was building more planes than the rest of the world put together.

In September 1947, the army lost control of its air force. The Department of Defense was formed and along with it, the United States Air Force. At that time the personnel strength had dropped from a wartime high of two million to just 300,000 men and women. At its transfer date, the Army Air Force had 25,000 planes, compared with nearly 70,000 at the end of the war two years earlier. Responsibility for both land and air forces was too much for one agency, and it was essential to cut aviation free and allow it to develop on its own.

Between 1908 and 1947 progress in military aviation was the primary responsibility of the army and today's success with advanced air force projects is a result of that initial development.

As an independent military unit, the U.S. Air Force continued to grow and now employs about 600,000 men and women. All its accomplishments since 1947 have rested on the foundation of its incredible record of achievement while under army supervision. The army, meanwhile, still uses planes, but only for general support duties. When helicopters proved their worth in Vietnam, they strengthened the role air units could play in ground warfare. Aircraft are once again a vital part of army operations.

THE MODERN AIR ARM

Today, the U.S. Army operates the largest number of military aircraft outside the Soviet Union. The army has a fleet of more than 9,000 aircraft, of which about 580 are fixed-wing planes and 8,600 are helicopters. The U.S. Air Force operates about 7,000 aircraft, and only 180 of those are helicopters. The U.S. Navy has a total of 2,300 aircraft, including 400 helicopters. The Air National Guard operates 1,800 aircraft, and the U.S. Marine Corp about 1,000.

Management of army aircraft is the responsibility of the Army Aviation System Command (**AVSCOM**), headquartered in St. Louis, Missouri. The command is also responsible for development of new aircraft and of systems for aircraft and helicopters. AVSCOM tries to improve the standard of equipment army aviators are asked to operate, and it increases the quality of that equipment where possible. This means that AVSCOM deals regularly with private companies and industrial corporations that get contracts to build army planes.

The modern army relies upon a wide range of land and air units operating collectively or individually as dictated by battlefield conditions.

AVSCOM employs more than 5,300 civilians in addition to 360 military personnel. It has an annual budget of around $6,000 million, most of which is spent with contractors on aircraft and flight equipment. AVSCOM also has a strong interest in research and development. Whenever technology opens an opportunity for better or more capable aircraft, AVSCOM puts people to work testing the new ideas.

AVSCOM buys all new equipment, including new aircraft and new systems that make the aircraft more effective. It makes sure that all the essential spare parts to keep the planes operating are available. AVSCOM is also responsible for deciding when aircraft should be retired from duty and how they should be disposed of — whether they should be broken up or have parts removed for use elsewhere.

AVSCOM's duties have been complicated by a new concept in military aviation, the **Airland Battle**. Previously, the air force operated independently of the army and the Marine Corps. Only when tasks required special services did one branch call upon the support of another. Of course, all land, sea, and air forces work cooperatively on the battlefield, but the lines of responsibility are clearly drawn.

Tanks and armored vehicles provide only limited protection for the crew because rockets and missiles fired from helicopters can penetrate even the most heavily armored vehicles. ▶

Tanks and other tracked vehicles moving across open land are an easy target for helicopters and light attack planes. ▼

13

▲
A close link between ground and air units has introduced the use of the AH-64 Apache attack helicopter, which has powerful weapons and complex electronic control equipment.

Infantry operations regularly call for helicopter support to move troops, support their ground activity with airborne fire ▶ power, or to evacuate wounded personnel.

The goal of Airland Battle is to confuse the enemy and cause him to fight in more than one direction at the same time. Military strategists believe that this will reduce the speed of an attack and make it less effective. Very large concentrations of armor and military hardware are centered in eastern Europe, and for that reason the U.S. Army believes that this is where it may face its biggest challenge yet.

Although highly unlikely, if Soviet armed forces ever broke through and launched an attack on western Europe, the U.S. Army would be at the forefront, trying to halt that assault and turn the enemy back toward his own territory. The United States is part of the **North Atlantic Treaty Organization (NATO)**. Formed in 1949, NATO is an alliance of 15 countries, including 11 European countries, Canada, Iceland, Great Britain, and the United States.

Because NATO countries are pledged to aid each other, an attack on one is an attack on all. The U.S. Army wants to be prepared to stop an attack should one ever occur. This does not mean that the army expects an attack from the Soviet Union or other non-NATO countries, but that it wants to be ready if one should occur.

Infantry units on the modern battlefield are moved from place to place by lightly armored personnel carriers and these, too, become objects of attack in war.

History shows that conflict can be avoided by achieving a level of readiness for war, because the aggressor sees how difficult his ambitions would be to put into practice. The United States adopts this policy of readiness and believes that through an alliance like NATO, the 15 countries will together pose such a level of force that the enemy will be deterred from attacking.

Careful study of Soviet military capabilities and of books and manuals on Soviet warfare have resulted in the concept of the Airland Battle, which the army supports. This strategy confronts the problem presented by the enormous increase in helicopters operated by armies throughout the world. In a battle, **anti-armor helicopters** would be used to attack tanks and armored columns. In retaliation, fighting helicopters would attack the enemy anti-armor helicopters. The Airland Battle strategy seeks to coordinate land and air forces in an integrated response.

In 1962, the army adopted the Airland Battle to reverse previous ways of stopping large tank and armored columns advancing at high speed. Instead of reacting to the enemy, the army would seize control of the battlefield by making the enemy respond to the army. It would do this by moving faster than the enemy, attacking to left and right of the battlefield and hitting him in the rear where his supplies come from.

Air force ground attack planes, such as the Republic Thunderbolt, would be used to hit large columns of tanks and to harass the enemy in every way possible. Army helicopter gunships would strike infantry units and **armored personnel carriers** as well as light tanks and artillery positions. In all aspects of the action, coordination of land and air forces would be essential. It is for that sort of defense that army aircraft are being prepared.

Careful design and construction has made the modern helicopter easier to service.

Vietnam was a proving ground for the concept of the helicopter gun ship which, back in the 1960s, was a helicopter with at least one gunner hanging over the side.

Troops must operate in areas where land vehicles and conventional aircraft cannot go and the CH-47 transport helicopter plays a vital role in keeping units supplied. ▼

SPIES IN THE SKY

To gather information about enemy positions and to spy on enemy radio conversation which may provide details about troop movements, the army uses the Grumman OV-1 Mohawk.

The army operates two types of piloted aircraft designed to spy on enemy activities by collecting radio signals and picking up communications. Information about the enemy's intentions is passed to commanders on the ground. This allows the army units to expect an activity they would not otherwise know about. That activity might be an attack, or it might be a large movement of troops and equipment through an unprotected area.

The planes that do this job are called Guardrail and Mohawk. Guardrail V is a Beechcraft RU-21H, and Improved Guardrail V is a Beechcraft RC-12D. Both types are versions of the basic Beechcroft C-12 light cargo plane. The army has about 170 Guardrails in service. They are 44 feet long with a wingspan of 54 feet and have a maximum weight of 7 tons. Cruising speed is 242 MPH and maximum range is approximately 1,700 miles. The aircraft entered service with the army in 1975.

Designed in the late 1950s, the Grumman OV-1 Mohawk has seen long service with the army as a **tactical reconnaissance** and **electronic intelligence-gathering plane**. Currently, the army operates about 100 for reconnaissance and 40 that perform a job similar to the Guardrail planes. Mohawk entered service in 1961 as a platform from which special sensors could observe the movements on the enemy on the battlefield. This is not an enviable job, because if detected, the plane would be attacked.

Now a separate military air arm, the U.S. Air Force helps the army by using planes like this OV-10 Bronco to spot the enemy's big guns and report back on the accuracy achieved by the army's howitzers and cannons.

Mohawk has a wingspan of 48 feet and a length of 45 feet, including a special radar pod beneath the forward fuselage designed to create radar "pictures" of enemy activity. Powered by two turboprop engines each rated at 1,400 horsepower, the plane has a speed of 290 MPH and a range of 944 miles. The Mohawk is quite vulnerable, and in future battle environments, unmanned **remotely piloted vehicles (RPVs)** would seem to be a better bet.

Remotely piloted vehicles, or RPVs, are unmanned model ◀ *aircraft remotely controlled from the ground.*

Although used primarily by the army, RPVs are also used by the navy; both services recover their robot planes by flying ▶ *them into safety nets.*

One version of the U.S. Air Force C-12 seen here is used by the army for spying on enemy activities. It can pick up radio signals over a long distance. ▼

RPVs are controlled by a "pilot" sitting on the ground. They are ideal for highly dangerous environments, such as the heat of a battlefield, or for spy missions over enemy territory. They also come in handy for locating artillery targets and for helping the guns monitor the accuracy with which they hit those targets. The RPV is very difficult to detect because of its miniature size, but development problems have not given the army great confidence, and several different types are still being looked at for production.

The Lockheed Aquila is a typical RPV. It is a swept-wing, tailless plane 5 feet long and 12 feet across. Aquila weighs 119 pounds at takeoff and can carry 36 pounds of sensors and electronic equipment. It has a maximum speed of 140 MPH and can remain airborne for about three hours. The Aquila carries a TV camera and can transmit information to the ground immediately or store it and send it later. It is controlled from the ground by a person using a small control stick. The stick is connected to a radio transmitter that sends the appropriate commands to fly the plane and control its 10-horsepower engine.

The RPV is launched along a rail and carries equipment which can take pictures of enemy activity or collect radio communications

Piloted planes and RPVs are unsuited for several jobs. These include long periods of spying and observing the enemy at relatively close range from a concealed or hidden position in all types of countryside. Helicopters are better at this type of duty, and the army has about 2,100 helicopters operating in this role. The oldest type still in use is the McDonnell Douglas OH-6A Cayuse, which first flew in 1963 and has been with the army since 1966. Almost all remaining Cayuse helicopters are with the Air National Guard or reserve units.

The Cayuse was named for an American range horse and is also the name of a Native American tribe. It is very lightweight and weighs little more than 1 ton fully loaded. Including rotors, it is 30 feet long, and it has a top speed of 150 MPH and a typical range of about 380 miles. The army has about 360 Cayuses, one of the most compact vehicles in aviation history. The standard Cayuse carries two crew members and four troops or up to 1,000 pounds of weapons including guns or **grenade launchers**.

The OH-6 Cayuse won the lightweight helicopter contest by beating a design from Bell eventually called the OH-58 Kiowa. Bell improved its contender, and the army

The OH-58 Kiowa was named after an Indian nation and has a top speed of 140 MPH with a maximum range of 300 miles.

decided to buy the Kiowa. Named after another Indian nation, Kiowa is classed as a light observation helicopter and in 1981 won a second competition arranged by the Army Helicopter Improvement Program (**AHIP**). The improvement program has been set up to improve an existing helicopter design and adapt it for a more specialized reconnaissance role.

The Kiowa weighs almost 3,000 pounds at takeoff and has a top speed of 140 MPH with a maximum range of 300 miles. The new AHIP version will have a more powerful engine, new rotor design, new and improved electronics for controlling the helicopter in flight, and a special TV camera on a mast above the rotors. This enables the TV to "see" the enemy with the helicopter hidden from view and gives it a big advantage when the enemy is watching for observation planes of all types.

▲
Rigged for a test involving surface-to-air missiles, this helicopter loses its tail with a direct hit just behind the engine.

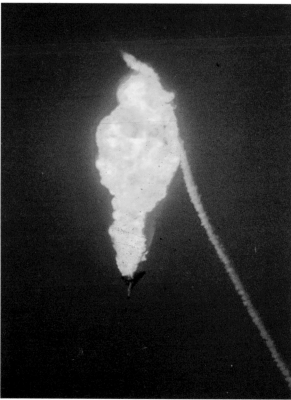

Enemy ground attack planes would seriously disrupt army ◄ *action, and air defense systems must be potent enough to destroy these planes too.*

The army uses a wide range of surface-to-air missiles to keep control of the skies over the battlefield; portable ► *shoulder-launched rounds provide infantrymen with a ready-made air defense system.*

The army currently has about 1,800 Kiowas and intends to convert almost 600 helicopters to the improved version by rebuilding them from original airframes. By 1988 less than 200 had been converted. The Kiowa is being operated with air cavalry, attack helicopter, and field artillery units as a support vehicle in the observation role. The modified version is very distinctive. Unlike the earlier model, it has angular windows that give it a wedge-shaped appearance. Earlier models had rounded windows that caused distortion and gave a false impression of what was happening on the ground below.

CARGO FLIGHT

The army operates about 500 transport helicopters. The most common is the Boeing Vertol CH-47 Chinook. Classed as a medium-lift helicopter, the Chinook has appeared in several different versions since its first flight in 1961. It has been built to carry cargo weights of up to 4,000 pounds stored inside, or to lift 16,000-pound weights from a hook and cable slung underneath. The heavier loads can include anything from ammunition and stores to light guns and artillery pieces.

Significant improvements made to the Chinook have greatly increased its performance. The latest model can lift up to 12 tons and carry cargo close to the forward edge of the battle for resupplying troops and artillery batteries. The helicopter is very reliable. For safety, it has two engines and two sets of rotors, each 60 feet in diameter. The CH-47 Chinook has a maximum takeoff weight of 25 tons.

The army operates more than 400 Chinooks, which are built to be as adaptable as possible. The Chinook normally has four wheel assemblies to achieve good

The standard all-purpose transport helicopter in service with the army today is the Boeing Vertol CH-47 Chinook which made its first flight almost thirty years ago.

The Chinook's design has been modified and improved over the years; present versions can lift up to 12 tons.

footing on rough or uneven ground. It can land on water, where the boat shape of its underbody allows it to float, and it can extend its normal operating range of 100 miles by taking on extra fuel in the air from a tanker helicopter.

The CH-47 is one of the rare government defense projects that continually stayed within price expectations. In all, Boeing Vertol built around 1,000 Chinooks, and more than 500 were operated in Vietnam. In addition to serving as a troop transporter and a supplies carrier, it was used for rescuing aircraft shot down by enemy action. More than 11,000 U.S. Air Force aircraft valued at more than $3,000 million were rescued by the Chinook's sky crane.

For really heavy jobs, the army falls back on the Sikorsky CH-54A Tarhe, a giant helicopter with a maximum loaded weight of more than 23 tons. It first flew in 1962. The Tarhe has no fuselage; instead a beam connects the front section where the crew sits with the tail rotor at the back. Including rotor blades, the Tarhe has a length of 88 feet, 6 inches and stands 18 feet, 7 inches high. Performance includes a maximum speed of 115 MPH, a cruising speed of 98 MPH, a range of 241 miles and a hover ceiling of 6,900 feet.

▲
The army operates a number of other support and transport helicopters such as the Sikorsky Black Hawk.

Transport helicopters must operate under all flying conditions and cross many different types of terrain to deliver supplies or recover wounded troops. ▶

The CH-54B is an improved model of the original Sikorsky Tarhe. It has performed exceptionally, lifting just over 20 tons and reaching a height of 36,122 feet. Not many duties call for it to lift such heavy cargo. Improvements were also made to the Chinook, enabling it to lift as much as the CH-54B but more efficiently. The Chinook is cheaper to run and more flexible to operate. Only about seventy Tarhes remain in service with the army, although a modification program will extend its life still further.

The army also operates a variety of fixed-wing aircraft for cargo duties, including the Beechcraft C-12 described in the preceding section. The army is prohibited from operating medium or heavy cargo planes, and the C-12 is basically a military version of the Beechcraft Super King Air 200 civilian commuter and light transport plane. It can carry urgent freight or a few passengers, and the army currently has about 98 of this type in three versions.

Three Black Hawk helicopters are delivered after their acceptance ceremony at the Sikorsky Aircraft Plant at Stratford, Connecticut.

ATTACK!

The present range of army helicopters in service owes its origin to early designs like this Type-47 Sioux which received fame through a TV series about helicopters.

The U.S. Army has two primary attack helicopters, the Bell AH-1 Cobra and the McDonnell Douglas AH-64 Apache. Both are potent representatives of the aggressive role helicopters can play if given the right design and the correct equipment. The army has about 1,400 attack helicopters, and more than 1,100 are Cobra versions that date back to 1963. In that year, Bell first flew a dramatically modified version of its well-known Type 47 Sioux helicopter of TV *Whirlybirds* fame.

Bell first built this helicopter as the Type 207, with a rounded glass nose and a gunner sitting alongside the pilot. Soon, though, Bell recognized the need for more power and a radically different approach to armed helicopter design. By the mid-1960s, the army was getting heavily involved in Vietnam and wanted an attack helicopter gunship that could fly and fight like a fixed-wing fighter. It had to be able to use forward-firing guns, launch rockets, and chase other helicopters.

To accomplish these jobs, in 1965 Bell came up with a modified version of the UH-1B Iroquois, a general-purpose helicopter with open sides and a couple of machine guns. In the modified form it appeared as the Type 209 HueyCobra, and the army immediately ordered 110. The orders never stopped coming and the company eventually built several different and improved versions. The latest is the AH-1S, which can carry a wide range of weapons to a maximum weight of 1.5 tons, an incredible capability for an agile, fighting helicopter.

The HueyCobra has a top speed of 170 MPH and a range of up to 350 miles, depending on the version and the weapons load carried. Up to 8 anti-tank guided missiles can be carried, with a 20mm gun in the nose turret and a 7.62mm gun alongside. A wide range of high explosives, rockets, grenade launchers, and missiles can be carried on short, stubby **winglets** on either side of the rear cockpit. The most recent version, the SuperCobra, has a top speed of 184 MPH, and the Marine Corps has designated this as its primary attack helicopter.

The Bell UH-1 Iroquois has had a long and distinguished career since it first saw service in Vietnam and became adopted by various units for many different roles.

The latest version of this famous Bell helicopter is the AH-1S which is a flexible, agile fighting helicopter that carries a wide range of rockets and missiles on its small winglets.

Air defense against enemy helicopters and other low-flying
aircraft is an important part of defense; this gatling gun carries
a series of rotating barrels, each firing in turn for very high
speed action.

The AH-1 is a light helicopter gun ship developed directly from
the UH-1 Huey. ▼

The army has started a "Cobra fleet life extension program," or **C-FLEX**. Engineers are making improvements to the rotor hub, the engine drive shaft, safety features, the cockpit, and radio communication equipment. The improved helicopters will have a higher reliability rate, and the army will operate them more efficiently and for less cost. Another program, the **C-NITE** upgrade, will give the Cobra a 24-hour tank-killing capability. This capability is achieved with special sights that allow the weapon-aiming equipment to "see" tanks day or night.

While members of the Cobra family are operated in larger numbers than any other army attack helicopter, the machine the enemy fears most is probably the McDonnell Douglas AH-64A Apache. The Apache has revolutionized the concept of the attack helicopter and forced several changes to Soviet tactics and equipment. Quite simply, the job of Apache is to destroy tanks — in very large numbers, in all types of weather, day or night, and over a wide range of countryside.

Apache is built to attack large concentrations of tanks and armor on the battlefield or to attack specific targets in open or concealed territory. To do this it carries two crewmen: one to fly the plane and the other to operate the complex weapons and their different aim and fire systems. Through special sensors and pieces of optical equipment, the target is located and its precise location stored on video tape.

The helicopter can rise from a forest clearing on a dark, moonless night, and special night sights allow its weapons officer to "see" the target and record it on video. Using a special flight control computer, the helicopter takes itself back down into the same hidden clearing even though the pilot cannot see a thing outside the window. The two crew members then replay the video, select the weapons they will use to destroy the target, and program the computer-controlled firing systems.

The next time the helicopter goes up above the trees, it wastes no time in locking onto the target and firing the appropriate weapons. The Hellfire missile is the most commonly used weapon for anti-tank work. It is a 2.75-inch rocket capable of punching a hole right through a tank turret. Several other types of missiles can be carried, and the Apache also has a chain gun of tremendous firepower. The gun, a 30mm Hughes M230A1, has a rapid rate of fire. It can discharge more than ten rounds per second, although short bursts would be used since the helicopter carries only 1,200 rounds.

A direct hit from an air defense system whose powerful radar and computers locate and track targets in the sky.

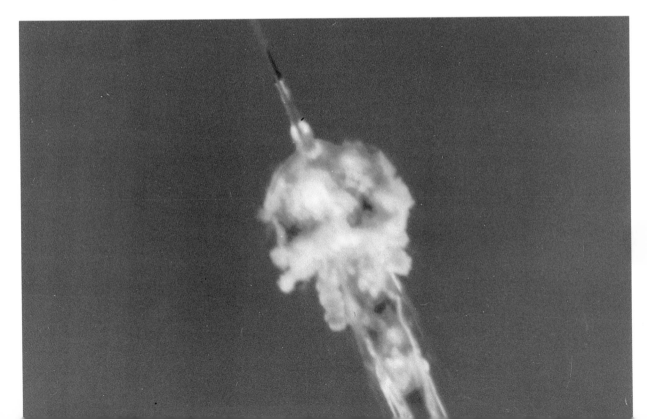

Sophisticated air defense systems are essential for major battle and combat operations where enemy planes must be driven away from sensitive areas.

The most potent heavy attack helicopter in the world today is the McDonnell Douglas AH-64A Apache which has a wide range of armament and an incredible performance unmatched by any other plane. ▼

Apache has a two-man crew — one crew member flies the plane and the other controls the weapons systems and search radars.

The AH-64A has a body length, including tail rotor, of 48 feet. The main four-blade rotor has a diameter of 28 feet. Maximum weight is almost 9 tons, and Apache has a top speed of 227 MPH and a creditable cruising speed of 182 MPH. The helicopter has an operating ceiling of more than 10,000 feet and a maximum range on internal fuel of 428 miles.

Apache is a very survivable helicopter to fly on attack missions. Its airframe is designed to survive fire from 12.7mm and 23mm guns. Two turboshaft engines each deliver almost 1,700 horsepower, making Apache reliable. Its performance is nothing less than breathtaking, and anyone who watches it cartwheel, roll, and loop through the sky gets a tremendous impression of its power and maneuverability.

Production of the Apache began in 1982, and by 1988 300 machines had been delivered to the army. The army wants about 675 Apache helicopters and expects to equip 34 battalions with it. The first battalion was fielded in May 1986, at Fort Hood, Texas. This helicopter is expected to operate with the army for many years into the future and will certainly be in front line service at the turn of the century.

The Apache's special sensors and optical equipment enable the weapons control system to seek out and lock onto several different targets simultaneously, select the appropriate ones to attack, and fire the weapons.

SUPPORT DUTIES

In addition to the attack and transport roles, the army uses rotary-winged planes to reach places its troops would otherwise be denied, such as cliff tops or rocky ledges.

Trucks, trailers, guns, and light artillery are all moved by air over difficult country.

By far the largest number of any single helicopter type operated by the army are the 3,500 Bell UH-1 Iroquois helicopters. These provide general support duties, which include troop movement, medical evacuation, supplies delivery, and training duties. The Iroquois helicopter goes back a long time; the first one made its initial flight in October 1956. More than thirty years later, it is the mainstay of the army's utility helicopter fleet.

The Iroquois formed the basis upon which Bell designed the successful Cobra family of attack helicopters. Since the army currently operates more than 1,100 Cobras, the fundamental design is maintained for more than 4,600 of the Army's 8,600 helicopters. The UH-1 gained great fame in Vietnam and became a popularly recognized shape from the TV series M★A★S★H, which was about medical teams in the Korean War.

The Iroquois is powered by a single **turboshaft engine** built by Lycoming. This engine has an output of between 600 horsepower and 1,400 horsepower. Later models are powered by a Pratt & Whitney turboshaft engine rated at 1,800 horsepower. Maximum speed is 127 MPH and range is typically 248 miles. The Iroquois weighs nearly 3 tons empty and can carry 2.5 tons of cargo or supplies. It has a length of 53 feet including rotors, and it lands on short skids, one on each side.

The army is replacing some of its Iroquois with a variety of other helicopters but expects to keep about 2,700 into the next century. By then, if still in service, it is possible that the Iroquois will have seen fifty years of production since the first machine rolled out the factory gates in 1958. By any standard, this is an astonishing feat for any aircraft, especially a helicopter.

In addition to the large fleet of Iroquois with medium lifting capacity, the army has about 1,000 UH-60 Black Hawk heavy-lift helicopters. First flown in 1974, Black Hawk is one of a family of highly versatile Hawk helicopters all based on the same design adapted for different roles. The U.S. Air Force and the U.S. Navy operate a large number of Night Hawk and Sea Hawk versions for a wide range of duties.

The army's Black Hawk is designed to work and survive close to hostile activity, delivering troops to and from the battlefield. It is also used to evacuate medical cases and lift guns and artillery pieces. Black Hawk can carry an eleven-man infantry squad or a 105mm howitzer with crew and ammunition. It can carry almost 5 tons above its empty weight of 5 tons with a range of 373 miles and a top speed of 184 MPH, making it ideal for troop operations. It can lift 4 tons from an underbody sling.

The Black Hawk is powered by two General Electric 1,600-horsepower turboshaft engines driving a four-blade rotor 53 feet, 8 inches in diameter. Two 7.62mm machine guns provide defense against attack. The army expects to take delivery of 1,107 helicopters of this type, and some are being modified for special duties. One, the EH-60, is to be used for detecting and jamming enemy radio signals. About 66 will be converted from the main production line of basic UH-60A types. Sikorsky, the manufacturer, is working on a developed version, the UH-60B, with improved performance and operational capabilities.

Heavy transport helicopters are frequently called upon to move trucks and supplies. Air force and army operations are ▶ coordinated to make the best use of the available equipment.

Small helicopters are built to fold down into a manageable size so they can be stowed aboard air force transport planes and moved quickly over long distances. ▼

A UH-60 Black Hawk lifts medical supplies on a mercy mission exercise.

THE FUTURE

Army interest in aviation has come full circle. From having established the great air forces upon which a completely separate United States Air Force was born in 1947, the army now has the largest number of military aircraft outside the Soviet Union. In addition, the army is looking to develop new aircraft and helicopters for the future. The army is taking on new responsibilities, and air support will always figure largely in decisions about the best equipment to have for the job at hand.

The army has decided to replace about 7,000 of the light helicopters currently in service, because they are getting old and obsolete. Its Light Helicopter Experimental (**LHX**) program is trying to develop a light, agile, twin-engine helicopter using advanced technology and the very latest materials that combine great strength with light weight.

The LHX helicopter will be expected to replace the AH-1, OH-58, OH-6, and UH-1 fleets while expanding the capabilities of the Black Hawk and Apache. The army plans to buy 4,500 LHX helicopters, the single largest order ever placed by the army. The LHX will be built in two versions. One will carry out the duties of a scout and light attack role, while the other will be a general-purpose helicopter, just like the types it is expected to replace.

Wherever military forces operate, helicopters will always play a vital role in rescuing wounded and injured troops. ▶

A UH-1D Iroquois passes over an M113 armored personnel carrier as a light infantry brigade prepares to move out. ▼

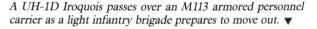

The LHX will need to be highly maneuverable and yet rugged, to survive in a battlefield environment where smoke and gunfire can quickly bring down the best helicopters around. It will be designed for air transportation and must be capable of being taken quickly in large planes to any place on earth. Once there, it must be able to fight combat missions against enemy helicopters or strike enemy ground forces. It must be capable of flying day or night and in adverse weather.

Perhaps future army air support requirements will call for a multi-purpose vehicle capable of flying into the battlefield area like a conventional fixed -wing aircraft and tilting its engine to become a helicopter for vertical descent as it nears the ground.

Several helicopter manufacturers have teamed together into two competing groups to bid for the big LHX contract. One team is led by Sikorsky and Boeing. The other is led by McDonnell Douglas and Bell. No matter which team wins the contest, the LHX will employ the very latest technology and the most recent developments in engines, rotors, materials, and design. It will weigh more than 4 tons loaded and cost no more than about $10 million for the scout version and $7 million for the general-purpose model.

Whatever the future holds for the LHX, the army has a very secure hold on its use of aircraft. Military operations on the battlefield would be unthinkable today without support from the air, and success in controlling the outcome of a battle rests largely on how well the aircraft and the airborne units are operating. The army knows this and has developed skills and capabilities that give troops on the ground the best possible chance of achieving their objectives with minimum force and minimum casualties.

The V-22 Osprey is a step in this direction, and may be the future shape of army air support.

Boeing's Heavy Lift Helicopter project was a proposed development of the CH-47 Chinook but work was stopped several years ago.

ABBREVIATIONS

AHIP   Army Helicopter Improvement Program

AVSCOM   Army Aviation Systems Command

C-FLEX   Cobra Fleet Life Extension

C-NITE   Cobra upgrade to night flight capability

LHX   Light Helicopter Experimental

GLOSSARY

Airland Battle	A concept developed by the army and the air force involving the cooperative use of aircraft and ground units in a combined activity.
Anti-armor helicopters	Helicopters equipped with guns and missiles for attacking heavily armored vehicles like tanks and armored personnel carriers.
Armored personnel carriers	Tracked or wheeled vehicles protected with armor plate and used to carry soldiers or infantry men.
Electronic intelligence-gathering plane	An aircraft that picks up electronic signals from enemy ground or air units in an attempt to gain advance information about their plans.
Grenade launcher	A device carried by helicopters for ejecting grenades at high speed onto ground targets.
North Atlantic Treaty Organization (NATO)	An alliance of the U.S., Canada, and 11 West European countries operating under a military pact to support one another; an attack on one is considered an attack on all.
Remotely Piloted Vehicle (RPV)	A pilotless aircraft about the size of a model plane that carries cameras or electronic sensors for spying or intelligence-gathering.
Tactical reconnaissance	Obtaining information that is of direct use to ground operations in battlefield action.
Turboshaft engine	A jet engine that receives its main thrust from a turbine-driven propeller.
Winglets	Short stubby fins placed on either side of a fuselage or cockpit with short under-winglet pylons for carrying rockets, grenade launchers, or missiles.

INDEX

Page references in *italics* indicate photographs or illustrations.

Airland Battle — 12, 15, 16
Allen, Brigadier-General James — 6
American Civil War — 6
Army Air Corps — 8, 10
Army Air Force — 10
Army Air Service — 8
Army Aviation System Command
    (AVSCOM) — 11, 12

B-29 Superfortress — *9*
Beechcraft
    C-12 — 18, *20-21*, 29
    RC-12D — 18
    RU-21H — 18
    Super King Air 200 — 29
Bell
    AH-1 Cobra — 30, *33*, 41
    AH-1S — 31, *32*
    Cobra helicopters — 34, 39
    OH-58 Kiowa — *23*, 24, 41
    SuperCobra — 31
    Type 47 Sioux — *30*
    Type 207 — 31
    Type 209 HueyCobra — 31
    UH-1 Iroquois — 39, 41
    UH-1B — *31*
    UH-1D — *42*
Boeing Vertol CH-47 Chinook — *17*, 26, 27, 29

General Electric — 40
Grumman OV-1 Mohawk — *18*, 19, 21
Guardrail — 18, 19
Guardrail V — 18

Hellfire missile — 34
Hughes M230A1 gun — 34

Improved Guardrail — 18

Light Helicopter Experimental
    (LHX) — 41, 42

Lockheed Aquila — 22
Lycoming — 39

McDonnell Douglas
    AH-64 Apache — *14*, 30, 34, *36*, *37*, 41
    AH-64A — 37
    OH-6 Cayuse — 23, 41
    OH-6A — 23

Night Hawk — 39
North Atlantic Treaty Organization
    (NATO) — 15, 15

OV-10 Bronco — *19*

Pratt & Whitney — 39

Remotely piloted vehicles (RPVs) — *20*, *21*, *22*, 23
Republic Thunderbolt — 16

Sea Hawk — 39
Sikorsky
    CH-54A Tarhe — 27
    CH-54B — 29
    EH-60 — 40
    UH-60 Black Hawk — *28*, 29. 39, 40, *41*
    UH-60A — 40
    UH-60B — 40

V-22 Osprey — *44*
Vietnam — 10, 27, 31, 39

Wilson, President Woodrow — 8
World War One — 8
World War Two — 8
Wright Brothers — 6
Wright Model A biplane — 6